To solve math problems,
what do you do?
Listen for words
that will give you the clues!

About how many
balls can there be?
More than one hundred?
What do you see?

A **lot**, or a **few**?
One thousand, or **less**?
Word clues can help you
to make a good **guess**!

One duck is big. Six ducklings are small.
Together, how many are there **in all**?

Together and **all**
are the clues you just heard.
Did they tell you to **add**
or **subtract** all the birds?

Six spotted puppies,
and one runs **away**.
Do you **add** or **subtract** it?
What would you say?

You know to **subtract** when you hear, day to day, words such as **minus**, one **less**, **take away**.

Three **pairs** of flowers—
red, yellow, and pink.
Can you **skip count** by twos?
What do you think?

When you count **pairs**, it can be more fun to count two by two, instead of by ones.

Six lollipops for the two boys to **share**—what can they do to make sure that it's fair?

The words **split** and **share** can help you decide that to break into groups, it's best to **divide**!

One lemonade
costs twenty-five cents.
After you **buy** it,
how much have you **spent**?

Whenever you **spend** or **buy** or **pay**,
do you add to your money,
or take some away?

Three friends pull each side
in a tug-of-war game.
If one boy goes **away**,
will the sides be the **same**?

If **another** girl comes out to play with her friends, how can the two sides be **equal** again?

How many **left**, if one balloon floats **away**?
Clues can help you with math every day!